Ticks

Jonathan Kravetz

The Rosen Publishing Group's
PowerKids Press™
New York

Published in 2006 by The Rosen Publishing Group, Inc.
29 East 21st Street, New York, NY 10010

First Edition

Editor: Jennifer Way
Book Design: Ginny Chu

Photo Credits: Cover, pp. 1, 5, 14 (inset) Scott Bauer, USDA ARS, www.insectimages.org; p. 6 © W. Perry Conway/Corbis; p. 9 Scott Camazine/Photo Researchers, Inc.; p. 9 (inset top and bottom) Mat Pound, USDA ARS, www.insectimages.org; pp. 10, 17 © Anthony Bannister, Gallo Images/Corbis; p. 13 © Bernie Lynch, closerviews.com; p. 14 Jim Occi, BugPics, www.insectimages.org; p. 18 Kenneth H. Thomas/Photo Researchers, Inc.; p. 21 Allen C. Steere, www.insectimages.org.

Library of Congress Cataloging-in-Publication Data

Kravetz, Jonathan.
Ticks / Jonathan Kravetz.— 1st ed.
p. cm. — (Gross bugs)
Includes index.
ISBN 1-4042-3046-7 (lib. bdg.)
1. Ticks—Juvenile literature. I. Title.

QL458.K73 2006
595.4'29—dc22

2004030596

Manufactured in the United States of America

CONTENTS

1. Bloodsuckers 4
2. Hosts and How Ticks Find Them 7
3. What Ticks Look Like 8
4. What Ticks Eat 11
5. Tick Eggs 12
6. Young Ticks 15
7. Adult Ticks 16
8. Mating and Laying Eggs 19
9. Diseases 20
10. How People Fight Ticks 22
Glossary 23
Index 24
Web Sites 24

Bloodsuckers

Ticks are eight-legged pests that suck blood. They are **parasites**, which means they feed on other animals. The animals that ticks feed on are called **hosts**. Ticks are harmful to people because they can spread diseases, or illnesses, like Lyme disease.

Ticks are **arachnids**, not insects. Other arachnids include spiders, scorpions, and mites. The ticks found in the United States are separated into two main **families**. They are the hard ticks, or Ixodidae, and the soft ticks, or Argasidae.

Although there are more than 800 **species** of ticks in the world, only about 100 carry diseases. There are about 90 species of ticks in the United States, 12 of which carry diseases. Some of the most common disease-carrying ticks are the wood tick, the dog tick, and the deer tick.

Ticks feed on many animals, including dogs, birds, lizards, cattle, and people. The main difference between arachnids, such as ticks, and insects is the shape of their body. Insects are separated into three sections, and arachnids are separated into two sections. This is a black-legged tick, which lives in the United States.

The tick in this picture (circled) is feeding on a toad. Ticks often have a hard time finding a new host. For this reason ticks can live a long time without eating. Some species of ticks can live for more than a year without feeding.

Hosts and How Ticks Find Them

Ticks are found everywhere in the world they can find hosts. Many hosts, such as deer, are found around water or in forests. While searching for a host, most hard ticks live on the ground in grassy meadows and woods. Soft ticks live in the nests and houses where their hosts live.

Most ticks spend most of their lives on or near the ground, waiting for hosts. Ticks cannot move quickly to catch onto their hosts. Instead they climb onto weeds or tall grass or up onto fences and buildings to wait for them. Searching for a host is called questing. A tick might wait only a week for a host, but it might also wait a year! Ticks sense **vibrations** from their hosts' movements. They can also sense odors, such as exhaled breath. Ticks attempt to snag onto hosts as they pass near the tick.

What Ticks Look Like

Ticks have flat, oval-shaped bodies. Adults have eight legs, but tick **larvae** have only six. Adult ticks are usually between 3/16 and 1 1/8 inches (5–29 mm) long and are brownish red. Males and females are usually different colors and sizes. Females are usually larger and a slightly brighter color than males.

Ticks use their mouthparts to grab onto hosts and to suck their blood. They use the **hypostome** to drink blood. It has many backward-curving teeth on it that help the tick hold onto the host. This is why ticks are hard to remove. At the end of the hypostome are **chelicerae**, which the tick uses to cut into its host. The **abdomen** of a tick swells as it drinks blood. Some ticks increase their size 20 to 50 times and their weight more than 200 times as they feed.

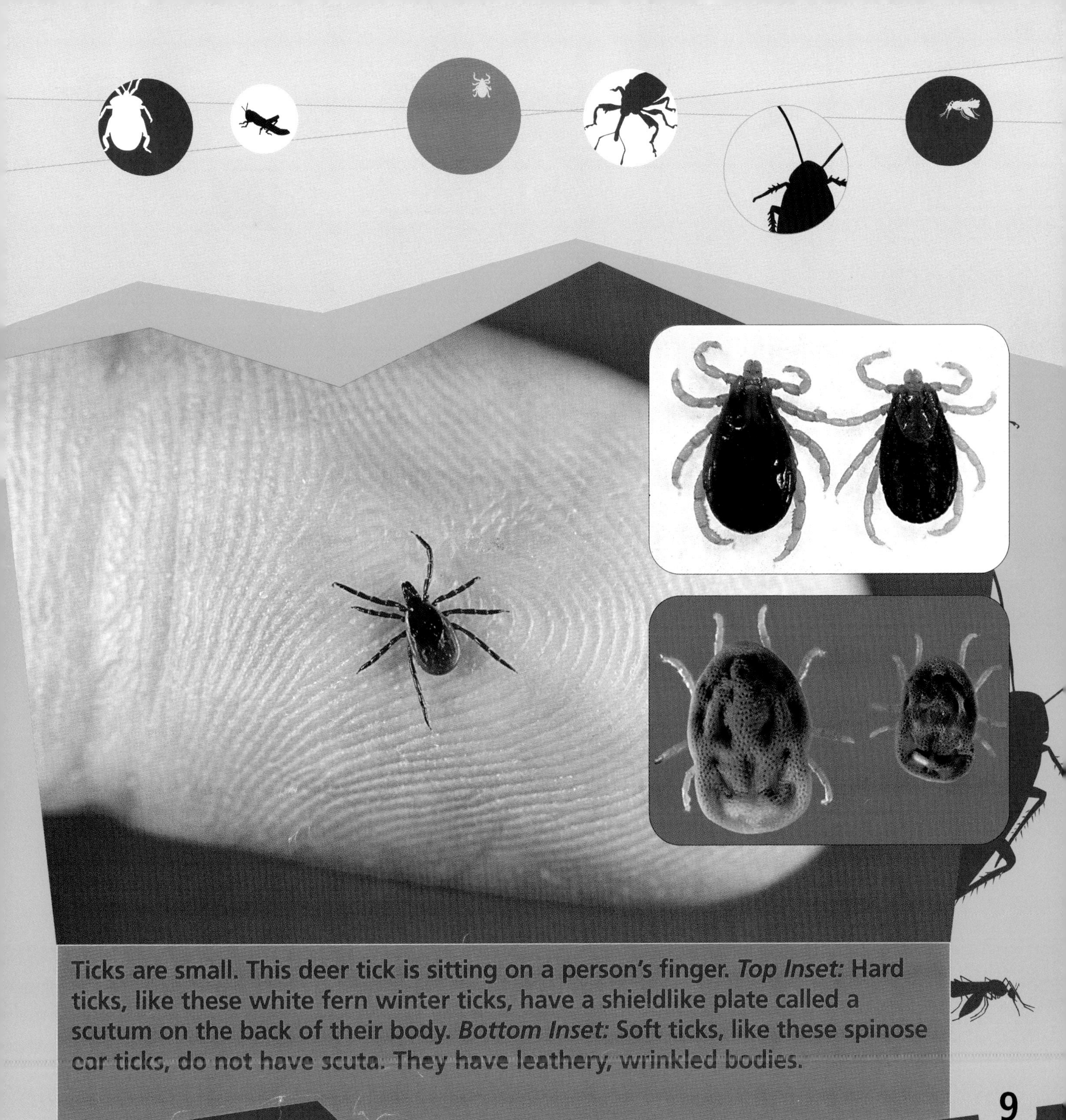

Ticks are small. This deer tick is sitting on a person's finger. *Top Inset:* Hard ticks, like these white fern winter ticks, have a shieldlike plate called a scutum on the back of their body. *Bottom Inset:* Soft ticks, like these spinose ear ticks, do not have scuta. They have leathery, wrinkled bodies.

The backward-curving teeth on a tick's hypostome keep the tick attached to its host while feeding, like the tick in this picture. Hard ticks even make a gluelike matter around the feeding area to attach more securely to their host. Ticks are hard to remove but must be removed properly to ensure that all of their parts have let go of the host.

What Ticks Eat

Ticks feed on blood. A few tick species feed on only one kind of host, but most species feed on many types of hosts.

Ticks attach themselves to a host's skin with their mouthparts. They use their chelicerae to cut an opening in the host's skin. The chelicerae cut blood vessels under the skin, which causes the blood to pool. The tick injects saliva, or spit, into the pooled blood. Ticks then suck this blood through the hypostome. While the tick is feeding, the hypostome holds the host so tightly that if you try to pull it off, the tick's body will rip off, leaving the mouth attached!

GROSS FACT

Ticks feed on mammals, birds, and even reptiles and amphibians. Ticks have many natural enemies, including birds, spiders, toads, fire ants, and centipedes. Certain wasp species lay their eggs inside tick nymphs. The hatching wasps then eat the tick nymphs from the inside out!

Tick Eggs

Ticks go through a four-stage life cycle. These stages are egg, larva, **nymph**, and adult. All stages except the egg stage require the tick to eat a blood meal from a host before changing into the next stage. The egg is the first stage of life.

Female soft ticks lay their eggs where the adult ticks live. They lay batches of 50 to 200 eggs. They feed after laying each batch. Hard ticks lay their eggs on the ground in a single mass that can have as many as 10,000 eggs. Eggs hatch in the spring when the weather begins getting warmer. Tick eggs usually hatch after one or two weeks. However, they can take as long as two months to hatch, depending on how warm it is.

This female hard tick is laying her eggs on the ground. After mating a female hard tick will feed on a host before laying one large batch of eggs, such as the one shown here. This batch can have as many as 10,000 eggs.

Ticks may live several months at the larval and nymphal stages. The length of time depends on the species. Ticks feed on hosts at both of these stages. The picture above shows the black-legged tick at three of its four life stages. Clockwise from the upper right corner is a larva, a nymph, an adult female, and an adult male. *Inset:* This is a black-legged tick nymph.

Young Ticks

The second and third stages of a tick's life are the larval and nymphal stages. Once the tick hatches from the egg, it is a larva. The six-legged larva feeds on a host. When it is done feeding, it drops to the ground, where it **molts**. Once the larva finishes molting, it has become a nymph. Nymphs are larger than larvae and have eight legs. Once nymphs find and feed on a host, they drop to the ground. There they molt to become adults. Adults are larger than nymphs and are able to **mate**.

There are a few differences between hard ticks and soft ticks at these stages. Many soft ticks go through several larval and nymphal molts, but hard ticks molt only once at each stage. Most hard ticks feed only once at each stage. Soft ticks feed several times during each stage.

Adult Ticks

After the nymph goes through its final molt and becomes an adult, it seeks out a host. Ticks can live through many poor conditions. They can live through weather that is very hot or very cold. Ticks can go a long time without feeding if they cannot find a host. Some soft ticks can live for more than a year without feeding.

The life cycle of a hard tick can last from less than a year to around three years. Soft ticks usually live longer than hard ticks. They can live for several years.

Ticks that feed on people, such as deer ticks, can be found near where people live. These ticks often remain hidden during the day and crawl out at night to find a host. After their first feeding, adult ticks are strong enough to mate.

Ticks spend most of their lives looking for a host. *Inset:* After it finds a host, the tick will attach itself and begin feeding. The tick in this picture has just attached itself to its host. As it feeds its body will swell with the host's blood.

These are adult male (left) and female (right) deer ticks. Hard ticks, such as these deer ticks, mate while on a host. After mating the male returns to the ground immediately, while the female feeds before she drops to the ground to lay her eggs.

Mating and Laying Eggs

To attract a mate, both female and male ticks give off a chemical called a **pheromone**. Pheromones help ticks find each other so they can mate.

Hard ticks usually mate on their hosts. After mating the male drops to the ground and soon dies. The female often stays on the host and feeds for several days. She then drops to the ground to lay her eggs. Female hard ticks feed only once after mating and lay one large batch of eggs. This batch often has as many as 10,000 eggs. Soon after laying her eggs, the female hard tick dies.

Soft ticks mate away from their host. The female lays many small batches of 20 to 50 eggs between feedings. Both male and female soft ticks usually live to mate more than once.

Diseases

Ticks do not really harm people by sucking their blood. Ticks cause the most harm by carrying and passing along many diseases, such as Lyme disease. Some ticks can also cause a condition called tick **paralysis**. Tick paralysis is caused by a poison that certain ticks inject into a person while feeding. The paralysis begins in the person's legs and arms and can even be deadly if the tick is not found and removed.

The most common disease caused by ticks in the United States is Lyme disease. The deer tick is the main carrier of Lyme disease. Deer ticks are most common in areas of the Northeast, the upper Midwest, and the Pacific coast. Drugs called antibiotics usually cure diseases caused by ticks, but only if they are caught early enough.

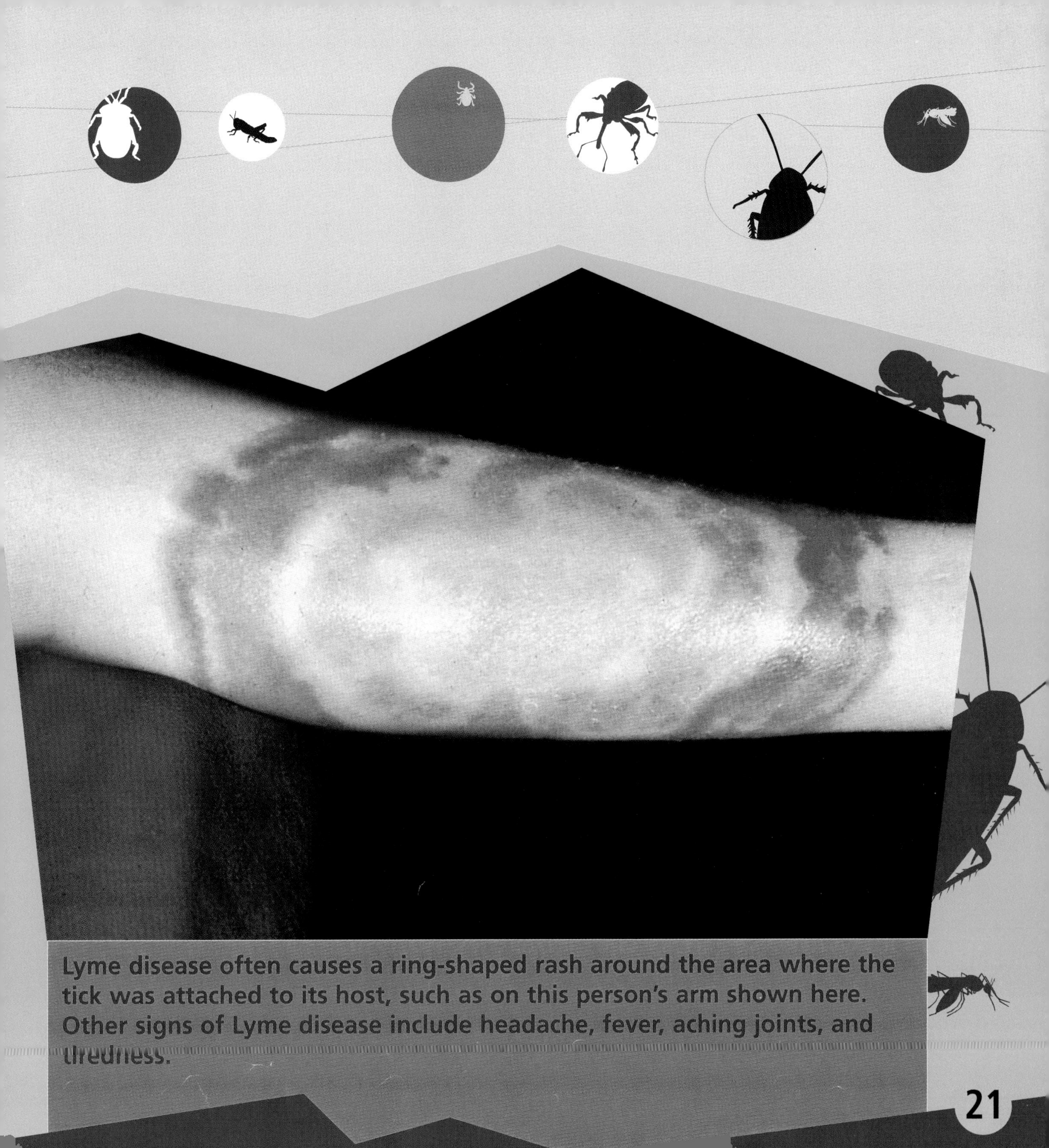

Lyme disease often causes a ring-shaped rash around the area where the tick was attached to its host, such as on this person's arm shown here. Other signs of Lyme disease include headache, fever, aching joints, and tiredness.

How People Fight Ticks

The best way to fight ticks is to wear protective clothing when you are in the woods. Long sleeves, hats, and long pants tucked into socks can keep ticks off your body when you are in an area where ticks are found. Another common way to fight ticks is to treat the places where they live with **pesticides**.

If a tick does attach itself to you, it is best to remove it as soon as possible. The sooner you find and remove a tick, the less chance you have of getting a disease.

When you remove a tick, it must be removed properly. Doctors have found the best way to remove a tick is to pull it out with tweezers as closely as possible to where the mouthparts enter the skin. Although they are tiny, ticks are strong creatures that can cause problems for their hosts.

GLOSSARY

abdomen (AB-duh-min) The large, rear part of an insect's or arachnid's body.
arachnids (uh-RAK-nidz) A type of animal that includes spiders and ticks.
chelicerae (kih-LIH-seh-ree) An arachnid's "jaws," used for holding.
families (FAM-leez) The scientific name for large groups of plants or animals that are alike in some ways.
hosts (HOHSTS) Creatures that provide food for a parasite.
hypostome (HY-puh-stohm) The rodlike mouthpart on a tick.
larvae (LAR-vee) Insects in the early life stage, in which they have a wormlike form.
mate (MAYT) To join together to make babies.
molts (MOHLTS) Sheds hair, feathers, shell, horns, or skin.
nymph (NIMF) A young insect that has not yet grown into an adult.
paralysis (puh-RA-luh-sus) Loss of feeling or movement in a part of the body.
parasites (PAR-uh-syts) Living things that live in, on, or with other living things.
pesticides (PES-tuh-sydz) Poisons used to kill pests.
pheromone (FER-uh-mohn) A kind of chemical produced by an animal that allows it to send a message to another of the same kind of animal.
species (SPEE-sheez) A single kind of living thing. All people are one species.
vibrations (vy-BRAY-shunz) Fast movements up and down or back and forth.

INDEX

A
arachnids, 4
Argasidae, 4

C
chelicerae, 8, 11
color, 8

D
deer tick(s), 4, 16, 20
dog tick, 4

H
host(s), 4, 7–8, 11–12, 15–16, 19, 22
hypostome, 8, 11

I
Ixodidae, 4

L
larva(e), 8, 12, 15
Lyme disease, 4, 20

M
molting, 15–16
mouthparts, 8, 11, 22

N
nymph(s), 12, 15–16

P
pheromone(s), 19

Q
questing, 7

S
size, 8

T
tick paralysis, 20

W
wood tick, 4

Web Sites

Due to the changing nature of Internet links, PowerKids Press has developed an online list of Web sites related to the subject of this book. This site is updated regularly. Please use this link to access the list:
www.powerkidslinks.com/gbugs/ticks/